PLANTS OF OKLAHOMA

An Herbarium of Roadside Flora

By

Eli Hartley

Contributions		Special Thanks
by	&	to
Tallulah Blue Hartley		Cieara Meyer & Cynthia Cole

Published by Phoenix Press & Kindle Publishing

Cole, Oklahoma

Contact: eli.l.hartley@gmail.com

(Also available as eBook)

First Edition – August 2022

Copyright © 2022 Eli Hartley

All rights reserved.

ISBN: 9798847252003

In the often-vaunted regions of the Far West, several hundred miles beyond the Mississippi, extends a vast tract of uninhabited country, where there is neither to be seen the log house of the white man, nor the wigwam of the Indian. It consists of great grassy plains, interspersed with forests and groves, and clumps of trees, and watered by the Arkansas, the grand Canadian, the Red River, and their- tributary streams. Over these fertile and verdant wastes still roam the elk, the buffalo, and the wild horse, in all their native freedom.

– Washington Irving,
Describing present day Oklahoma in *A Tour on the Prairies, 1835*

Contents

Introduction	i
Herbarium	1
References	103
Index	104
Notes/Leaf Press	105

Introduction

I. About

  This project started as a simple leaf collection intended for my daughter, who at the time of writing this is two years old. We live in a small town on a small farm in McClain County, situated in Central Oklahoma. I have always had a particular fondness for nature and the outdoors and have for as long as I can remember wanted to be able to identify every plant on our property. This desire grew into the desire for me to be able to hopefully teach my daughter as she grows to understand and appreciate the natural world around her. I wanted her to have a resource—a guide that would help her learn to identify the plants she will regularly encounter in our environment. From this it seemed like a natural progression to compile my work into a book and extend this knowledge to as many people as possible.

  Part of that progression also meant expanding the collection beyond our small farm. I wanted to include a variety of plants that Oklahomans, or visitors to our state might encounter. "Roadside flora", is just another way to say that the plants included are common, and accessible. You won't have to fight (too many) ticks or chiggers, to find these plants, because more often than not, they are found by your mailbox, in an empty lot, or the field down the road. The plants in this book are not rare, nor do they require an expedition to collect. These plants are available to those of us who have a keen eye, and a keen interest for the nature world, and I have found all of the plants in this book within a stone's throw from a road. I always keep a small hand shovel and pair of pruning shears in the car, and when I see something new or interesting (and safely accessible), I collect it.

  It should be noted that "roadside" could mean different things to different people. A cityscape is likely going to offer different plant choices than the rural roadside. Likewise, far Eastern Oklahoma will have a different selection than will the Western part of the state. Seasonal changes affect selections as well. What is found in summer will be different than winter. Sometimes days or weeks can even make all the difference. Taking these things into consideration, this is more accurately a book of roadside flora, found mostly in rural areas of Central Oklahoma, exclusively during summertime, during a drought. This name didn't quite have the same ring to it, so I landed where I did.

Photo (Left): Black Walnut (Juglans nigra) by Charles C. Deam, 1922

II. Herbarium

Before I started this project, I only had a vague idea of what an herbarium was. I knew that I wanted to assemble and prepare a collection of local plants but wasn't quite sure how to approach this task. I certainly had no idea that enormous collections already existed in dedicated departments at universities and museums across the world. Often these collections, (usually consisting of thousands of specimens) are assembled into collective and regional databases. These herbaria (multiple herbariums) are invaluable resources for student and scientist alike. Many of the world's herbaria have been around for hundreds of years. In fact, specimens collected on Darwin's voyage on the HMS Beagle can still be found intact and tucked safely away in a special cabinet somewhere.

As one might expect from a science steeped in so much history, there is a standard criterion that has emerged for compiling such a collection. I have bent the rules some to suit my needs but should you one day find yourself irresistibly compelled to create a collection of your own, it will be useful to have a fundamental awareness of the basics. Luckily, there's not a lot to it. Between this book, a quick web search, and some craft supplies, you should have all of the tools you will need. In its most simple form, an herbarium is nothing more than a leaf or a plant that has been pressed and dried in an old book, identified, and taped or pasted to a sheet of paper. Some of the earliest and most interesting herbaria I have seen have been nothing more than small journals, often with multiple specimens attached to a single page accompanied with handwritten notes.

If you aspire to create a modern university or museum quality specimen, however, you will need a bit more. The standard size for a paper mount is 11 ½ inches wide x 16 ½ inches long. The paper should be cotton rag, acid-free, and archival quality. The specimen should likewise be mounted using archival quality glue, (usually Elmer's), and have a 4x4 inch card that provides details such as the scientific name, common name, taxonomic family, specimen conditions when collected, date and location, and who the specimen was collected by. This info card is then glued to the bottom right corner of the herbarium sheet.

Special care is also taken when collecting plant specimens. The most useful specimens contain as much identifying content as possible. Ideally, this means including features such as the roots, flowers, fruit, seeds, stems, leaves, and anything else that belongs with the plant. Understandably, this may not always be possible, but it's worth aspiring towards. The specimen once collected needs to find its way to a plant press as soon as possible to minimize the wilt and degradation that will begin to occur as soon as it has been harvested. Before being placed in your press, which could be anything from an old book to a proper plant press with adjustable straps, care should be taken to clean the plant of dirt or debris, prune the specimen to fit your mount and carefully position the specimen for drying, as this will ultimately be its final form. When situating the plant specimen, be sure to have at least one leaf situated so that the back of the leaf is visible. Also be mindful to situate it in a natural and aesthetically pleasing way.

To dry, the plant is usually placed in between old newspaper print, and further sandwiched between "blotter paper" which is very similar in structure to the mounting paper. All temporary field notes or identifying cards are placed in the press along with the plant and the press is strapped down tight. After a few days, the blotter paper may become moist and need to be swapped out for dry paper. The blotter paper can be reused many times. It can take a few days to many weeks to dry depending on environmental conditions and the moisture content of the plant.

Once you have a dry plant, it's time to mount. There are many ways to do this, but I used a small paintbrush. I added a little water to my glue to thin it out and painted the back of the specimen with glue. Do the same to the back of your printed 4x4 card with all of the identifying info. Cover the mount with wax paper and place it underneath something that provides even pressure while drying. I would sandwich my mounts between old records, but bags of rice, or cardboard, and some old, stacked books would work too. After another day, the mount should be dry. If you have large stems or branches, or loose material you may need to further secure these by using tape or sewing the stems to the paper. Some plants just mount more attractively than others, but if you have taken this much time and effort to mount a specimen you will have a work of science and work of art that you will no doubt be proud of. It's also a creation that if properly stored and cared for could survive for hundreds of years or longer.

III. My Process

As mentioned above, all the plants in the book were collected by my daughter and I near a road. We did our best to collect specimens that best show the attributes of the plant, but due to the season (summer), we were not able to find fruit or flowers for some of the plants. The process for building my collection was simple and varied slightly from what is listed in the section above. I used 9x12 inch paper instead, because I knew this book would be 8.5x11 and I wanted the pages to be of similar size. I used acid free, 140lb rag paper used for water coloring found at a local big-box hobby store. While not perfect, it has worked well for my purposes. To dry the plants, I used printer paper and newsprint when I could find it. I also used square cutouts of cardboard to separate specimens. All my specimens were sandwiched in between two cut pieces of plywood, (one on top, one on bottom) with one or sometime two 25lb weights on top.

To identify plants, I used a variety of resources. I have a few go-to books, as well as encyclopedia's, online resources, forums, and herbaria, Google Lens, and the help of residents or botanists. I have a reference page in the back of the book that has all the resources I used. The Oklahoma Forestry Service (OFS) along with other state agencies and universities also always seem eager to help if you reach out. Finally, while brief, I have also included details with each specimen such as where it was found, soil or light conditions, flower color, or other unique characteristics about the plant.

IV. Final Thoughts

Creating an herbarium has been a rewarding process for many reasons. It has been a beautiful balance between art and science and has renewed my awareness and appreciation for the natural world around me. A task as mundane as driving has now become a treasure hunt, as I am always looking for the next new and interesting plant. The once blurry landscape has become clearer now that I can distinctly identify and understand many of the plants around me. Whether you intend to build a collection of your own, or just wish to better understand the rich plant life Oklahoma has to offer, I hope this book finds you well and helps you on your way.

25	**High Plains**		**33**	**East Central Texas Plains**
25b	Rolling Sand Plains		33a	Northern Post Oak Savanna
25c	Moderate Relief Plains		**35**	**South Central Plains**
25e	Canadian/Cimarron High Plains		35b	Floodplains and Low Terraces
26	**Southwestern Tablelands**		35c	Pleistocene Fluvial Terraces
26a	Canadian/Cimarron Breaks		35d	Cretaceous Dissected Uplands
26b	Flat Tablelands and Valleys		35g	Red River Bottomlands
26c	Caprock Canyons, Badlands, and Breaks		35h	Blackland Prairie
26f	Mesa de Maya/Black Mesa		**36**	**Ouachita Mountains**
27	**Central Great Plains**		36a	Athens Plateau
27d	Prairie Tableland		36c	Central Mountain Ranges
27h	Red Prairie		36d	Fourche Mountains
27i	Broken Red Plains		36e	Western Ouachitas
27k	Wichita Mountains		36f	Western Ouachita Valleys
27l	Pleistocene Sand Dunes		**37**	**Arkansas Valley**
27m	Red River Tablelands		37a	Scattered High Ridges and Mountains
27n	Gypsum Hills		37b	Arkansas River Floodplain
27o	Cross Timbers Transition		37d	Arkansas Valley Plains
27p	Salt Plains		37e	Lower Canadian Hills
27q	Rolling Red Hills		**38**	**Boston Mountains**
27r	Limestone Hills		38b	Lower Boston Mountains
28	**Flint Hills**		**39**	**Ozark Highlands**
28a	Flint Hills		39a	Springfield Plateau
29	**Cross Timbers**		39b	Dissected Springfield Plateau–Elk River Hills
29a	Northern Cross Timbers		**40**	**Central Irregular Plains**
29b	Eastern Cross Timbers		40b	Osage Cuestas
29c	Western Cross Timbers		40d	Cherokee Plains
29d	Grand Prairie			
29g	Arbuckle Uplift			
29h	Northwestern Cross Timbers			
29i	Arbuckle Mountains			

—— Level III ecoregion boundary
······ County boundary
—— Level IV ecoregion boundary
- - - State boundary

SCALE 1:1 250 000

Albers Equal Area Projection
Standard Parallels 34° 30' N and 36° 00' N

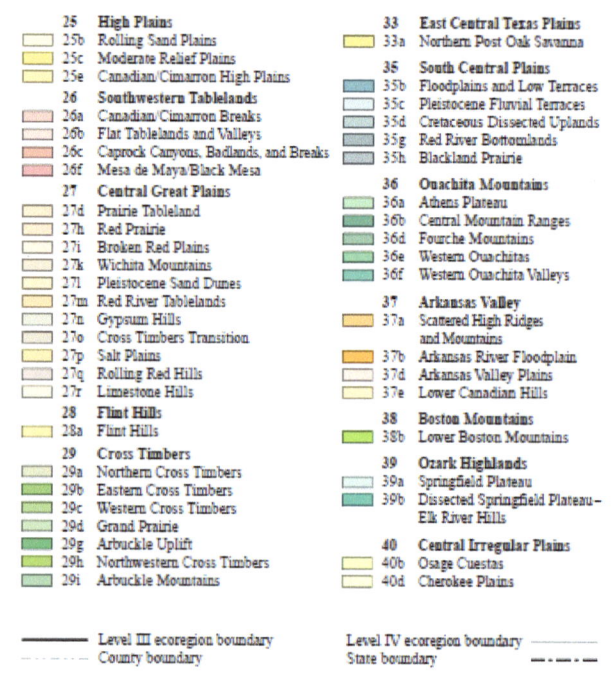

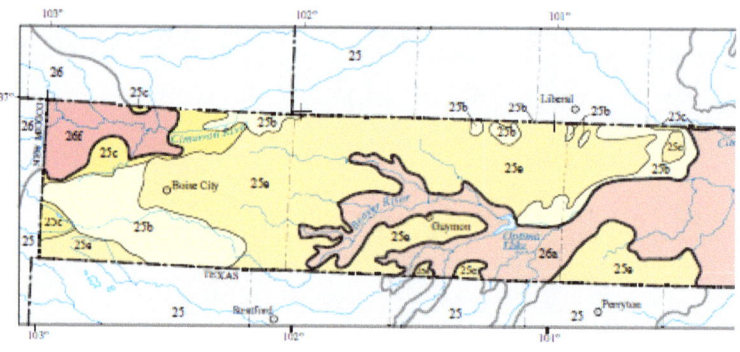

Level III Ecoregions of the Conterminous United States

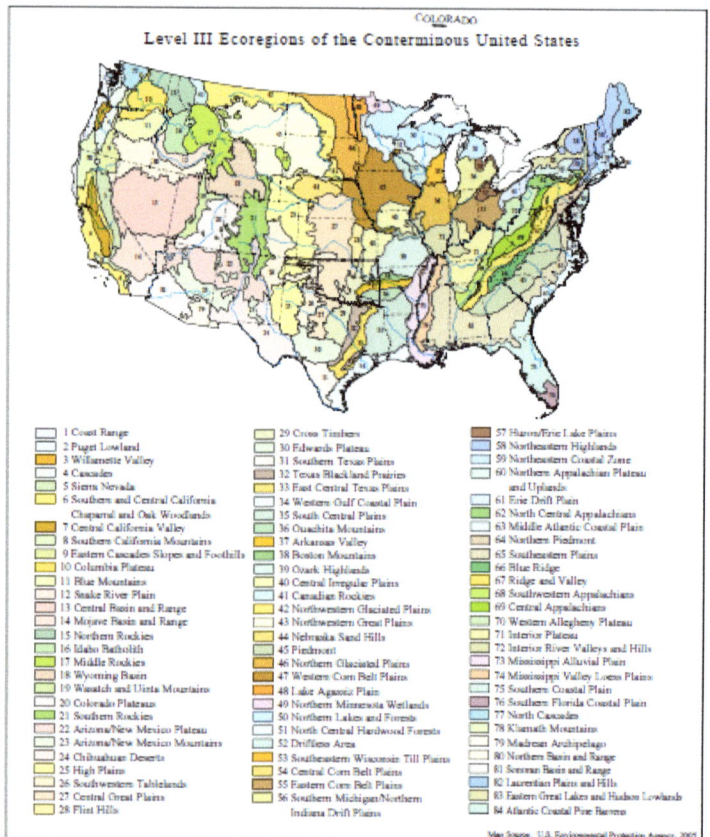

PRINCIPAL AUTHORS: Alan J. Woods (Oregon State University), James M. Omernik (U.S. Geological Survey), Daniel R. Butler (Oklahoma Conservation Commission–Water Quality Division), Jimmy G. Ford (U.S. Department of Agriculture–Natural Resources Conservation Service), James E. Henley (U.S. Department of Agriculture–Natural Resources Conservation Service), Bruce W. Hoagland (Oklahoma Biological Survey), Derek S. Arndt (Oklahoma Climatological Survey), and Brian C. Moran (Indus Corporation).

COLLABORATORS AND CONTRIBUTORS: Kurt Atkinson (Oklahoma Department of Agriculture, Food, and Forestry), Sandy A. Bryce (Dynamac Corporation), Shannen S. Chapman (Dynamac Corporation), Philip A. Crocker (U.S. Environmental Protection Agency), Glenn E. Griffith (Dynamac Corporation), Chris Hise (The Nature Conservancy), Charlie Howell (U.S. Environmental Protection Agency), Ron Jarman (Apex Environmental, Inc.), Thomas R. Loveland (U.S. Geological Survey), Kenneth V. Luza (Oklahoma Geological Survey), Phillip Moershel (Oklahoma Water Resources Board), Mark E. Moseley (U.S. Department of Agriculture–Natural Resources Conservation Service), Randy Parham (Oklahoma Department of Environmental Quality), and Brooks Tramell (Oklahoma Conservation Commission–Water Quality Division).

REVIEWERS: George A. Bukenhofer (U.S. Forest Service), Richard A. Marston (Boone Pickens School of Geology, Oklahoma State University), David V. Peck (U.S. Environmental Protection Agency), and Dale Splinter (Boone Pickens School of Geology, Oklahoma State University).

CITING THIS POSTER: Woods, A.J., Omernik, J.M., Butler, D.R., Ford, J.G., Henley, J.E., Hoagland, B.W., Arndt, D.S., and Moran, B.C., 2005, Ecoregions of Oklahoma (color poster with map, descriptive text, summary tables, and photographs): Reston, Virginia, U.S. Geological Survey (map scale 1:1,250,000).

This project was supported in part by funds from USEPA Region 6, Water Quality Cooperative Agreement under the provisions of Section 104(b)(3) of the Clean Water Act to the Oklahoma Water Resources Board (through the Office of the Secretary of Environment, State of Oklahoma). Assistance from the private sector is acknowledged in the form of Ron Jarman, Ph.D., on loan from Apex Environmental, Inc.

Electronic versions of ecoregion maps and posters as well as other ecoregion resources are available at http://www.epa.gov/wed/pages/ecoregions.htm

Ecoregions of Oklahoma

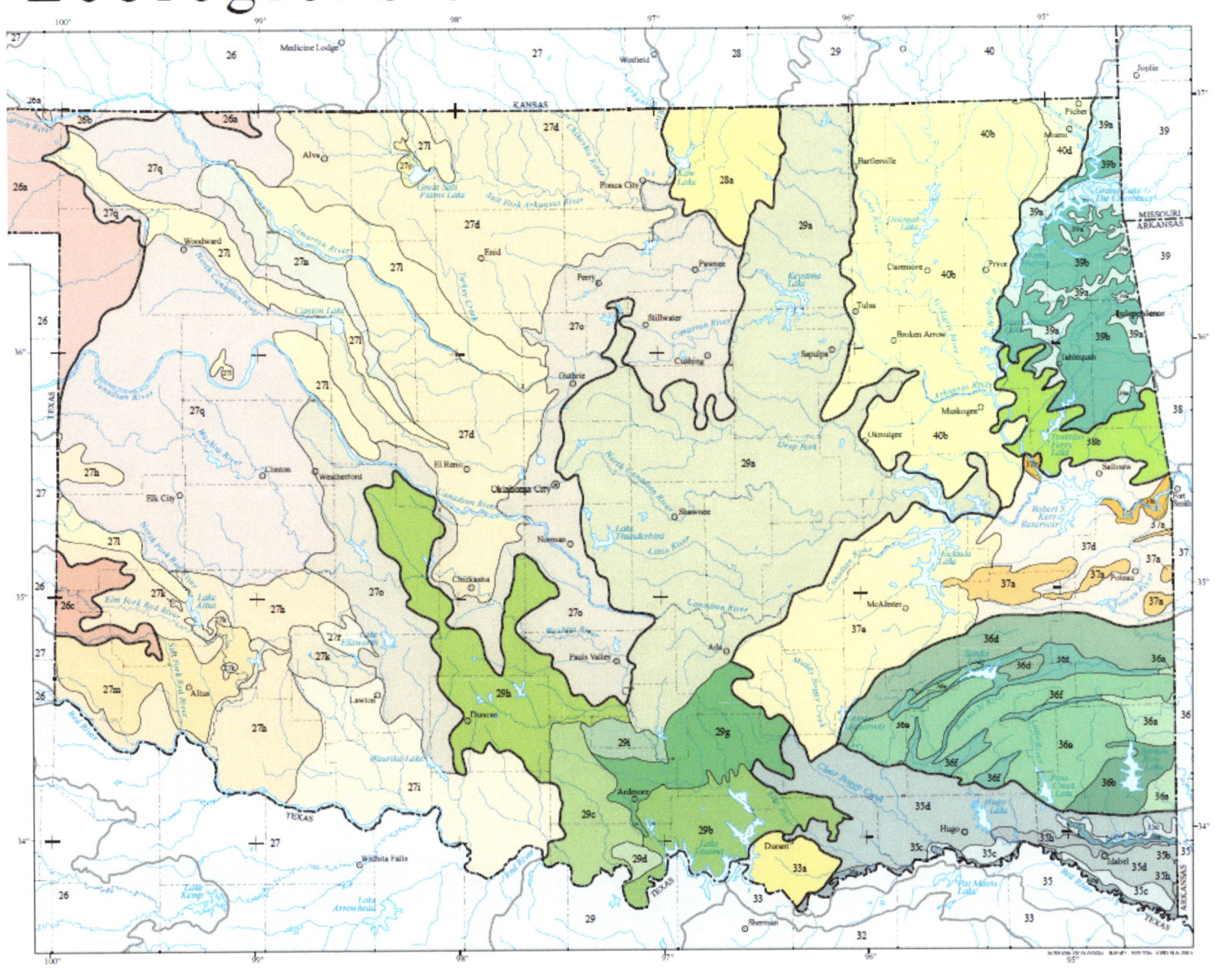

PLANTS OF OKLAHOMA

AN HERBARIUM OF ROADSIDE FLORA

Eastern Red Cedar
(Juniperus virginiana)

Plants of Oklahoma

Entry: 1 Date: 6-30-22
Scientific Name: Juniperus virginiana
Common Name: Eastern Red Cedar
Family: Cupressaceae
Location: Cole, OK- S. of Hwy 74 on May
Notes: Coniferous, evergreen, edible berries
Collected By: E. Hartley & T.B. Hartley

Bald Cypress
(Taxodium distichum)

Plants of Oklahoma

Entry: 37 Date: 7-10-22
Scientific Name: Taxodium distichum
Common Name: Bald Cypress
Family: Cupressaceae
Location: Lindsey St. Norman, OK
Notes: Deciduous, conifer growing in lot
Collected By: E. Hartley & T.B. Hartley

# Shortleaf Pine
(Pinus echinata)

Plants of Oklahoma

Entry: 83 Date: 7-18-22
Scientific Name: Pinus echinata
Common Name: Shortleaf Pine
Family: Pinaceae
Location: 240th & Penn, Cole, OK
Notes: Evergreen, coniferous, sandy soil
Collected By: E. Hartley & T.B. Hartley

Post Oak
(Quercus stellata)

Plants of Oklahoma

Entry: 3　　　　　　　　　Date: 6-30-22

Scientific Name: Quercus stellata

Common Name: Post Oak

Family: Fagaceae

Location: Cole, OK- S. of Hwy 74 on May

Notes: Deciduous, found in full sun, dry

Collected By: E. Hartley & T.B. Hartley

Blackjack Oak
(Quercus marilandica)

Plants of Oklahoma

Entry: 2   Date: 6-30-22
Scientific Name: Quercus marilandica
Common Name: Blackjack Oak
Family: Fagaceae
Location: Cole, OK- S. of Hwy 74 on May
Notes: Deciduous, found in dry sandy soil
Collected By: E. Hartley & T.B. Hartley

Dwarf Chinkapin Oak
(Quercus prinoides)

Plants of Oklahoma

Entry: 95　　　Date: 7-24-22
Scientific Name: Quercus prinoides
Common Name: Dwarf Chinkapin Oak
Family: Fagaceae
Location: 240th & Penn, Cole, OK
Notes: Deciduous tree/shrub, understory
Collected By: E. Hartley & T.B. Hartley

Sawtooth Oak
(Quercus acutissima)

Plants of Oklahoma

Entry: 105 Date: 8-6-22
Scientific Name: Quercus acutissima
Common Name: Sawtooth Oak
Family: Fagaceae
Location: Earlywine Park, Moore, OK
Notes: Deciduous tree, dry loam, full sun
Collected By: E. Hartley & T.B. Hartley

Shumard Oak
(Quercus shumardii)

Plants of Oklahoma

Entry: 91   Date: 7-25-22
Scientific Name: Quercus shumardii
Common Name: Shumard oak
Family: Fagaceae
Location: HWY 39 & Rockwell, Dibble, OK
Notes: Deciduous tree, large acorns, dry soil
Collected By: E. Hartley & T.B. Hartley

Water Oak
(Quercus nigra)

Plants of Oklahoma	
Entry: 98	Date: 7-20-22
Scientific Name:	Quercus nigra
Common Name:	Water Oak
Family:	Fagaceae
Location:	Earlywine Park, Moore, OK
Notes:	Deciduous tree, native, loamy soil
Collected By:	E. Hartley & T.B. Hartley

Sycamore
(Platanus occidentalis)

Plants of Oklahoma

Entry: 9 Date: 6-30-22
Scientific Name: Platanus occidentalis
Common Name: American Sycamore
Family: Platanaceae
Location: Cole, OK- S. of Hwy 74 on May
Notes: Deciduous, mature tree, sandy soil
Collected By: E. Hartley & T.B. Hartley

Sugarberry
(Celtis laevigata)

Plants of Oklahoma

Entry: 96 Date: 7-26-22
Scientific Name: Celtis laevigata
Common Name: Sugarberry
Family: Cannabaceae
Location: E Sooner Rd, Bridge Creek, OK
Notes: Deciduous tree, edible pea size fruit
Collected By: E. Hartley & T.B. Hartley

Chinese Elm
(Ulmus parvifolia)

Plants of Oklahoma

Entry: 103   Date: 8-6-22
Scientific Name: Ulmus parvifolia
Common Name: Chinese Elm
Family: Ulmaceae
Location: Earlywine Park, Moore, OK
Notes: Deciduous tree, dry loam, full sun
Collected By: E. Hartley & T.B. Hartley

Tree of Heaven
(Ailanthus altissima)

Plants of Oklahoma

Entry: 12 Date: 6-30-22

Scientific Name: Ailanthus altissima

Common Name: Tree of Heaven

Family: Simaroubaceae

Location: Cole, OK- S. of Hwy 74 on May

Notes: Deciduous, invasive, found in shade

Collected By: E. Hartley & T.B. Hartley

Southern Magnolia
(Magnolia grandiflora)

Plants of Oklahoma

Entry: 22   Date: 7-4-22

Scientific Name: Magnolia grandiflora

Common Name: Southern Magnolia

Family: Magnoliaceae

Location: Downtown Purcell, OK

Notes: Evergreen, growing in fertile soil

Collected By: E. Hartley & T.B. Hartley

Empress Tree
(Paulownia tomentosa)

Plants of Oklahoma

Entry: 32 Date: 7-8-22
Scientific Name: Paulownia tomentosa
Common Name: Empress Tree
Family: Paulowniacea
Location: Forestry Dept., Goldsby, OK
Notes: Annual, produces familiar stickers
Collected By: E. Hartley & T.B. Hartley

American Persimmon
(Diospyros virginiana)

Plants of Oklahoma

Entry: 90 Date: 7-26-22

Scientific Name: Diospyros virginiana

Common Name: American Persimmon

Family: Ebenaceae

Location: Walnut Creek Bridge, Washington, OK

Notes: Deciduous tree, edible fruit, sandy soil

Collected By: E. Hartley & T.B. Hartley

Elderberry
(Sambucus canadensis)

Plants of Oklahoma

Entry: 30 Date: 7-5-22
Scientific Name: Sambucus canadensis
Common Name: Elderberry
Family: Adoxaceae
Location: Cole, OK- S. of Hwy 74 on May
Notes: Deciduous, shrub, ripe berries purple
Collected By: E. Hartley & T.B. Hartley

Crape Myrtle
(Lagerstroemia indica)

Plants of Oklahoma

Entry: 88 Date: 7-16-22
Scientific Name: Lagerstroemia indica
Common Name: Crape Myrtle
Family: Lythraceae
Location: Cole, OK- S. of Hwy 74 on May
Notes: Deciduous tree/shrub, colorful flowers
Collected By: E. Hartley & T.B. Hartley

Gum Bully
(Sideroxylon lanuginosum)

Plants of Oklahoma

Entry: 16
Date: 7-3-22
Scientific Name: Sideroxylon lanuginosum
Common Name: Black Haw, Gum Bully
Family: Sapotaceae
Location: Cole, OK- S. of Hwy 74 on May
Notes: Deciduous, small shrub-like tree
Collected By: E. Hartley & T.B. Hartley

Holly
(Ilex aquifolium)

Plants of Oklahoma

Entry: 27 Date: 7-4-22
Scientific Name: Ilex aquifolium
Common Name: Holly
Family: Aquifoliaceae
Location: Canadian St. Purcell, OK
Notes: Evergreen, shrub, invasive
Collected By: E. Hartley & T.B. Hartley

Roughleaf Dogwood
(Cornus drummondii)

Plants of Oklahoma

Entry: 84   Date: 7-19-22
Scientific Name: Cornus drummondii
Common Name: Roughleaf Dogwood
Family: Cornaceae
Location: Walnut Creek Bridge, Washington, OK
Notes: Deciduous, tree/shrub, white flowers
Collected By: E. Hartley & T.B. Hartley

Curly Dock
(Rumex crispus)

Plants of Oklahoma

Entry: 21　　　　　　　　　Date: 7-4-22

Scientific Name: Rumex crispus

Common Name: Curly Dock

Family: Polygonaceae

Location: Cole, OK- S. of Hwy 74 on May

Notes: Perennial, invasive, growing in gravel

Collected By: E. Hartley & T.B. Hartley

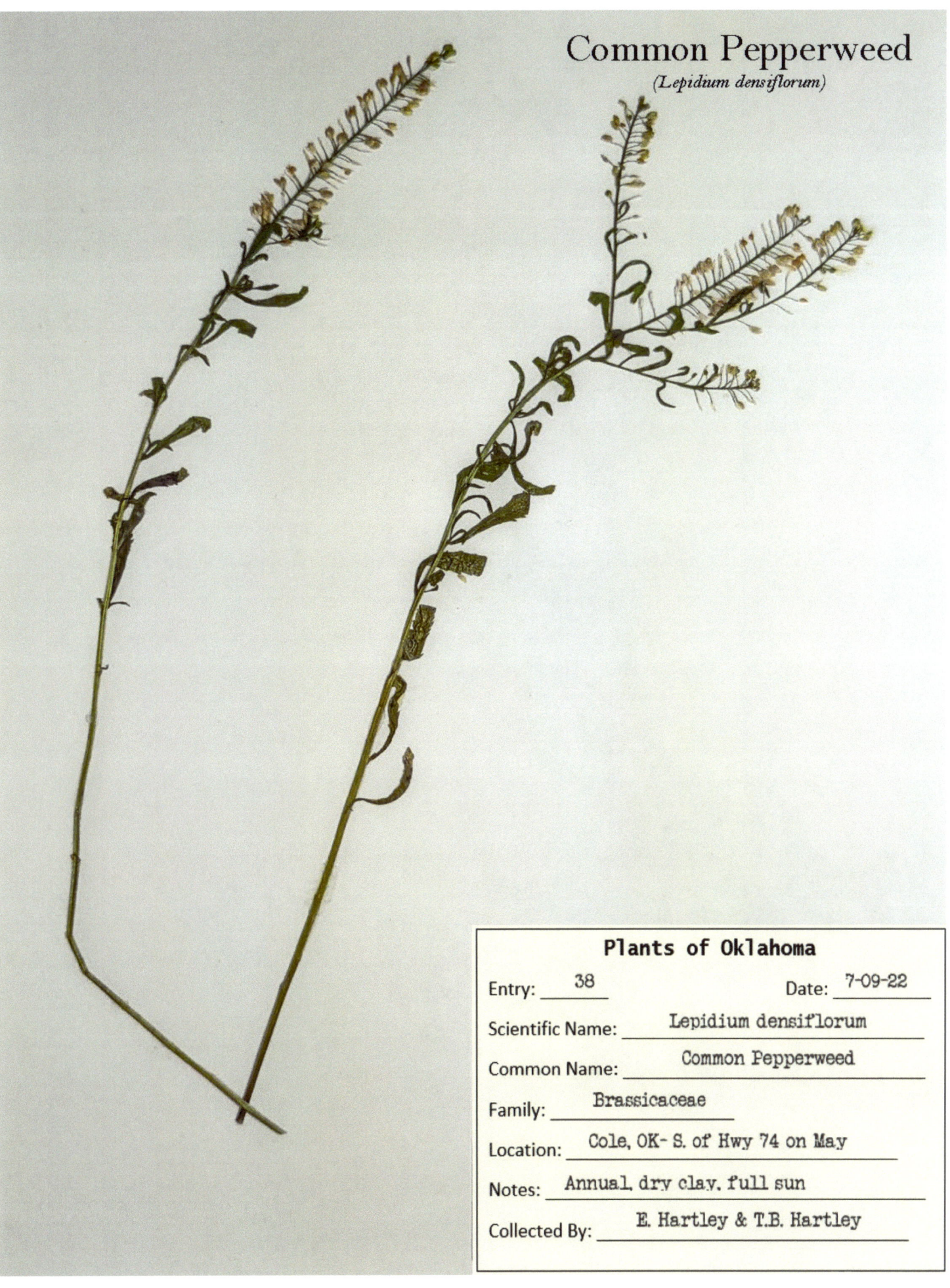

Common Pepperweed
(Lepidium densiflorum)

White Prickly Poppy
(Argemone albiflora)

Plants of Oklahoma

Entry: 53 Date: 7-15-22
Scientific Name: Argemone albiflora
Common Name: White Prickly Poppy
Family: Papaveraceae
Location: 240th & Penn, Cole, OK
Notes: Annual, flower, sandy dry soil
Collected By: E. Hartley & T.B. Hartley

Buffalo Gourd
(Cucurbita foetidissima)

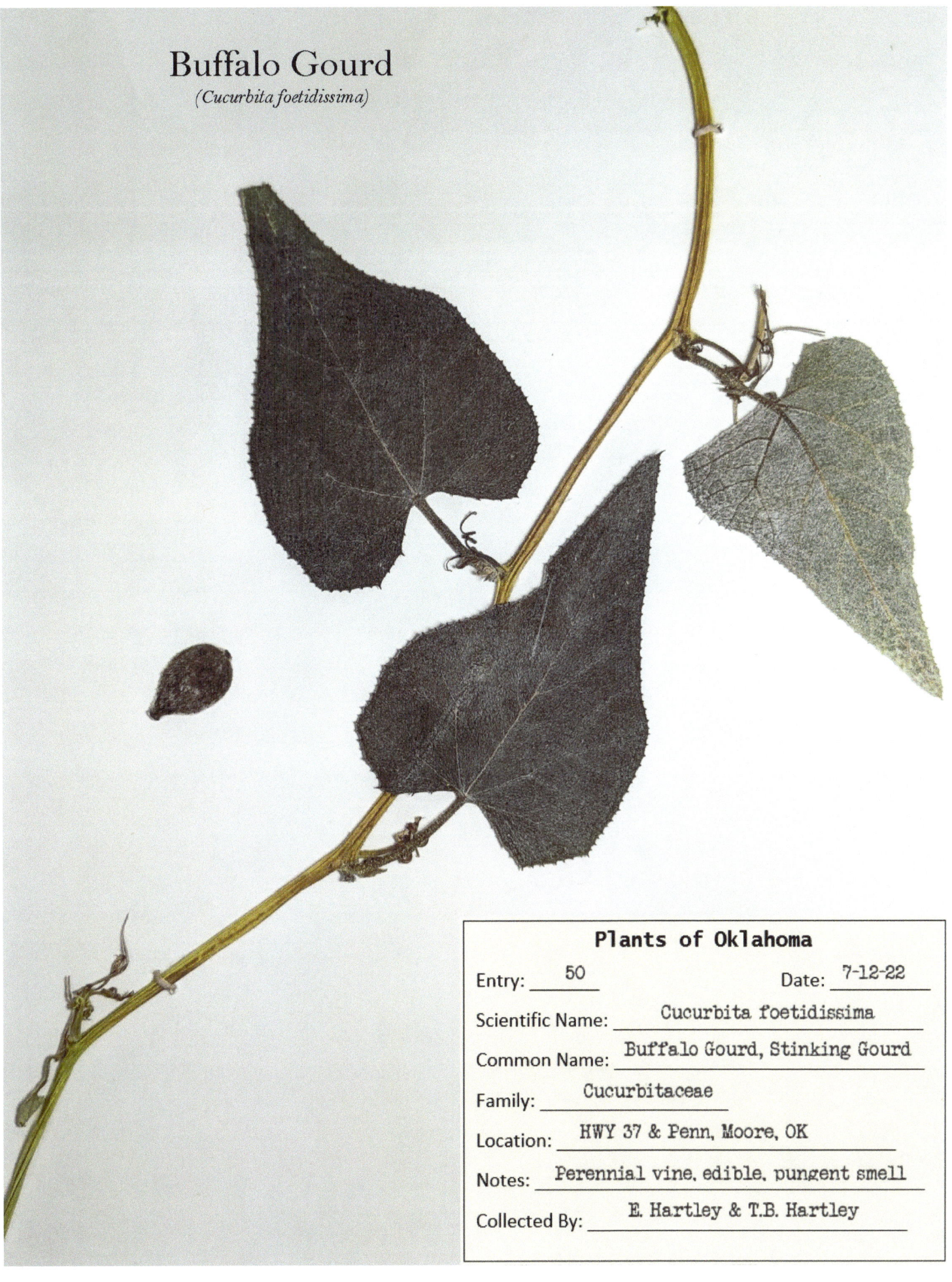

Plants of Oklahoma

Entry: 50 Date: 7-12-22
Scientific Name: Cucurbita foetidissima
Common Name: Buffalo Gourd, Stinking Gourd
Family: Cucurbitaceae
Location: HWY 37 & Penn, Moore, OK
Notes: Perennial vine, edible, pungent smell
Collected By: E. Hartley & T.B. Hartley

Pokeweed
(Phytolacca americana)

Plants of Oklahoma	
Entry: 45	Date: 7-12-22
Scientific Name:	Phytolacca americana
Common Name:	Pokeweed, Poke Sallet
Family:	Phytolaccaceae
Location:	May & HWY 39 Dibble, OK
Notes:	Perennial, toxic, red/purple berries
Collected By:	E. Hartley & T.B. Hartley

Roundleaf Greenbriar
(Smilax rotundifolia)

Plants of Oklahoma

Entry: 44 Date: 7-12-22
Scientific Name: Smilax rotundifolia
Common Name: Roundleaf Greenbriar
Family: Smilacaceae
Location: Cole, OK- S. of Hwy 74 on May
Notes: Woody vine, purple berries, edible
Collected By: E. Hartley & T.B. Hartley

Hoary Vervain
(Verbena stricta)

Plants of Oklahoma

Entry: 55 Date: 7-16-22
Scientific Name: Verbena stricta
Common Name: Hoary Vervain
Family: Verbenaceae
Location: I-35 at Goldsby, OK
Notes: Annual, dry sandy soil, purple flower
Collected By: E. Hartley & T.B. Hartley

Yellow Nutsedge
(Cyperus esculentus)

Plants of Oklahoma

Entry: 34
Date: 7-4-22
Scientific Name: Cyperus esculentus
Common Name: Yellow Nutsedge
Family: Cyperaceae
Location: Cole, OK- S. of Hwy 74 on May
Notes: Perennial, growing in full sun, dry
Collected By: E. Hartley & T.B. Hartley

Field Bindweed
(Convolvulus arvensis)

Plants of Oklahoma

Entry: 82 Date: 7-16-22
Scientific Name: Convolvulus arvensis
Common Name: Field Bindweed
Family: Convolvulaceae
Location: Cole, OK- S. of Hwy 74 on May
Notes: Perennial vine, rhizomes, dry loam
Collected By: E. Hartley & T.B. Hartley

Great Mullein
(Verbascum thapsus)

Plants of Oklahoma

Entry: 85   Date: 7-20-22
Scientific Name: Verbascum thapsus
Common Name: Great Mullein
Family: Scrophulariaceae
Location: HWY 39 & Rockwell, Dibble, OK
Notes: Biennial, hairy, dry soil, full sun
Collected By: E. Hartley & T.B. Hartley

Lemon Beebalm
(Monarda citriodora)

Plants of Oklahoma

Entry: 60 Date: 7-15-22
Scientific Name: Monarda citriodora
Common Name: Lemon Beebalm
Family: Lamiaceae
Location: Walnut Creek Bridge, Washington, OK
Notes: Annual, edible (mint family), sandy loam
Collected By: E. Hartley & T.B. Hartley

Honeysuckle
(Lonicera periclymenum)

Plants of Oklahoma

Entry: 64 Date: 7-16-22
Scientific Name: Lonicera periclymenum
Common Name: Honeysuckle
Family: Caprifoliaceae
Location: Cole, OK- S. of Hwy 74 on May
Notes: Perennial, deciduous vine, understory
Collected By: E. Hartley & T.B. Hartley

Palmer's Amaranth
(Amaranthus palmeri)

Plants of Oklahoma

Entry: 79 Date: 7-17-22
Scientific Name: Amaranthus palmeri
Common Name: Palmer's amaranth, Pigweed
Family: Amaranthaceae
Location: HWY 39 & Rockwell, Dibble, OK
Notes: Annual, edible, roadside, full sun
Collected By: E. Hartley & T.B. Hartley

Chinese Privet
(Ligustrum lucidum)

Plants of Oklahoma

Entry: 17　　　　　　　　Date: 7-3-22
Scientific Name: Ligustrum lucidum
Common Name: Chinese Privet
Family: Oleaceae
Location: Cole, OK- S. of Hwy 74 on May
Notes: Evergreen, invasive, shrub-like tree
Collected By: E. Hartley & T.B. Hartley

Japanese Privet
(Ligustrum japonicum)

Plants of Oklahoma

Entry: 26 Date: 7-5-22
Scientific Name: Ligustrum japonicum
Common Name: Japanese Privet
Family: Oleaceae
Location: Canadian St. Purcell, OK
Notes: Evergreen, shrub, invasive
Collected By: E. Hartley & T.B. Hartley

Eastern Poison Ivy
(Toxicodendron radicans)

Plants of Oklahoma

Entry: 42 Date: 7-5-22
Scientific Name: Toxicodendron radicans
Common Name: Eastern Poison Ivy
Family: Anacardiaceae
Location: Cole, OK- S. of Hwy 74 on May
Notes: Perennial, toxic, woody vine or shrub
Collected By: E. Hartley & T.B. Hartley

Common Milkweed
(Asclepias syriaca)

Plants of Oklahoma

Entry: 72 Date: 7-19-22
Scientific Name: Asclepias syriaca
Common Name: Common Milkweed
Family: Apocynaceae
Location: Walnut Creek Bridge, Washington, OK
Notes: Perennial, dry sandy loam, full sun
Collected By: E. Hartley & T.B. Hartley

Butterfly Weed
(Asclepias tuberosa)

Plants of Oklahoma

Entry: 69 Date: 7-15-22
Scientific Name: Asclepias tuberosa
Common Name: Butterfly Weed
Family: Apocynaceae
Location: HWY 62 & Main, Blanchard, OK
Notes: Perennial, native, dry sandy soil
Collected By: E. Hartley & T.B. Hartley

Box Elder Maple
(Acer Negundo)

Plants of Oklahoma

Entry: 66 Date: 7-19-22
Scientific Name: Acer negundo
Common Name: Box Elder (Maple)
Family: Sapindaceae
Location: Walnut Creek Bridge, Washington, OK
Notes: Deciduous, dry sandy loam, full sun
Collected By: E. Hartley & T.B. Hartley

Red Maple
(Acer rubrum)

Plants of Oklahoma

Entry: 24　　　　　　　　　Date: 7-3-22
Scientific Name: Acer rubrum
Common Name: Red Maple
Family: Sapindaceae
Location: Downtown Purcell, OK
Notes: Deciduous, growing in fertile soil
Collected By: E. Hartley & T.B. Hartley

Silver Maple
(Acer saccharinum)

Plants of Oklahoma

Entry: 8 Date: 6-30-22
Scientific Name: Acer saccharinum
Common Name: Silver Maple
Family: Sapindaceae
Location: Cole, OK- S. of Hwy 74 on May
Notes: Deciduous tree, edible sap, full sun
Collected By: E. Hartley & T.B. Hartley

Cottonwood
(Populus deltoides)

Plants of Oklahoma

Entry: 11 Date: 6-30-22
Scientific Name: Populus deltoides
Common Name: Eastern Cottonwood
Family: Salicaceae
Location: Cole, OK- S. of Hwy 74 on May
Notes: Deciduous, distinctively noisy in wind
Collected By: E. Hartley & T.B. Hartley

Black Willow
(Salix nigra)

Plants of Oklahoma

Entry: 7 Date: 6-30-22
Scientific Name: Salix nigra
Common Name: Black Willow
Family: Salicaceae
Location: Cole, OK- S. of Hwy 74 on May
Notes: Deciduous, found near pond
Collected By: E. Hartley & T.B. Hartley

Pecan
(Carya illinoinensis)

Plants of Oklahoma

Entry: 87 Date: 7-15-22

Scientific Name: Carya illinoinensis

Common Name: Pecan

Family: Juglandaceae

Location: HWY 77 & Maguire, Noble, OK

Notes: Deciduous, edible nuts, full sun

Collected By: E. Hartley & T.B. Hartley

Carolina Horsenettle
(Solanum carolinense)

Plants of Oklahoma

Entry: 35 Date: 7-5-22
Scientific Name: Solanum carolinense
Common Name: Carolina Horsenettle
Family: Solanaceae
Location: Hwy 9, Newcastle, OK
Notes: Perennial, toxic, dry sandy soil
Collected By: E. Hartley & T.B. Hartley

Buffalo Bur
(Solanum rostratum)

Plants of Oklahoma

Entry: 49 Date: 7-12-22

Scientific Name: Solanum rostratum

Common Name: Buffalo-Bur, Spiny Nightshade

Family: Solanaceae

Location: I-35 at Goldsby, OK

Notes: Annual, dry sandy soil, yellow flower

Collected By: E. Hartley & T.B. Hartley

Graybark Grape
(Vitis cinerea)

Plants of Oklahoma

Entry: 29 Date: 7-5-22
Scientific Name: Vitis cinerea
Common Name: Graybark Grape, Winter Grape
Family: Vitaceae
Location: Cole, OK- S. of Hwy 74 on May
Notes: Perennial, growing in understory
Collected By: E. Hartley & T.B. Hartley

Heartleaf Peppervine
(Ampelopsis cordata)

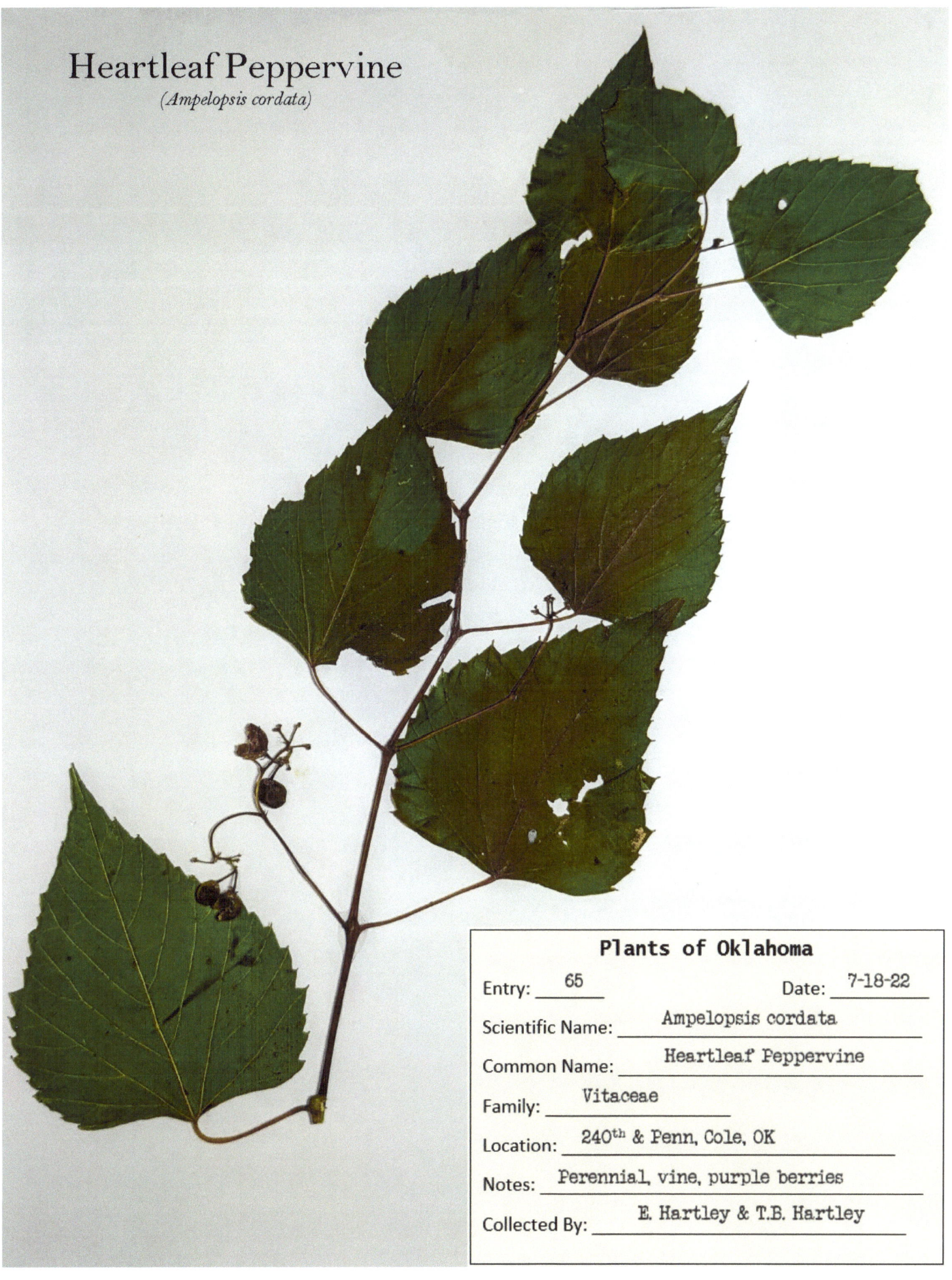

Plants of Oklahoma

Entry: 65 Date: 7-18-22
Scientific Name: Ampelopsis cordata
Common Name: Heartleaf Peppervine
Family: Vitaceae
Location: 240th & Penn, Cole, OK
Notes: Perennial, vine, purple berries
Collected By: E. Hartley & T.B. Hartley

Paper Mulberry
(Brossonetia papyrifera)

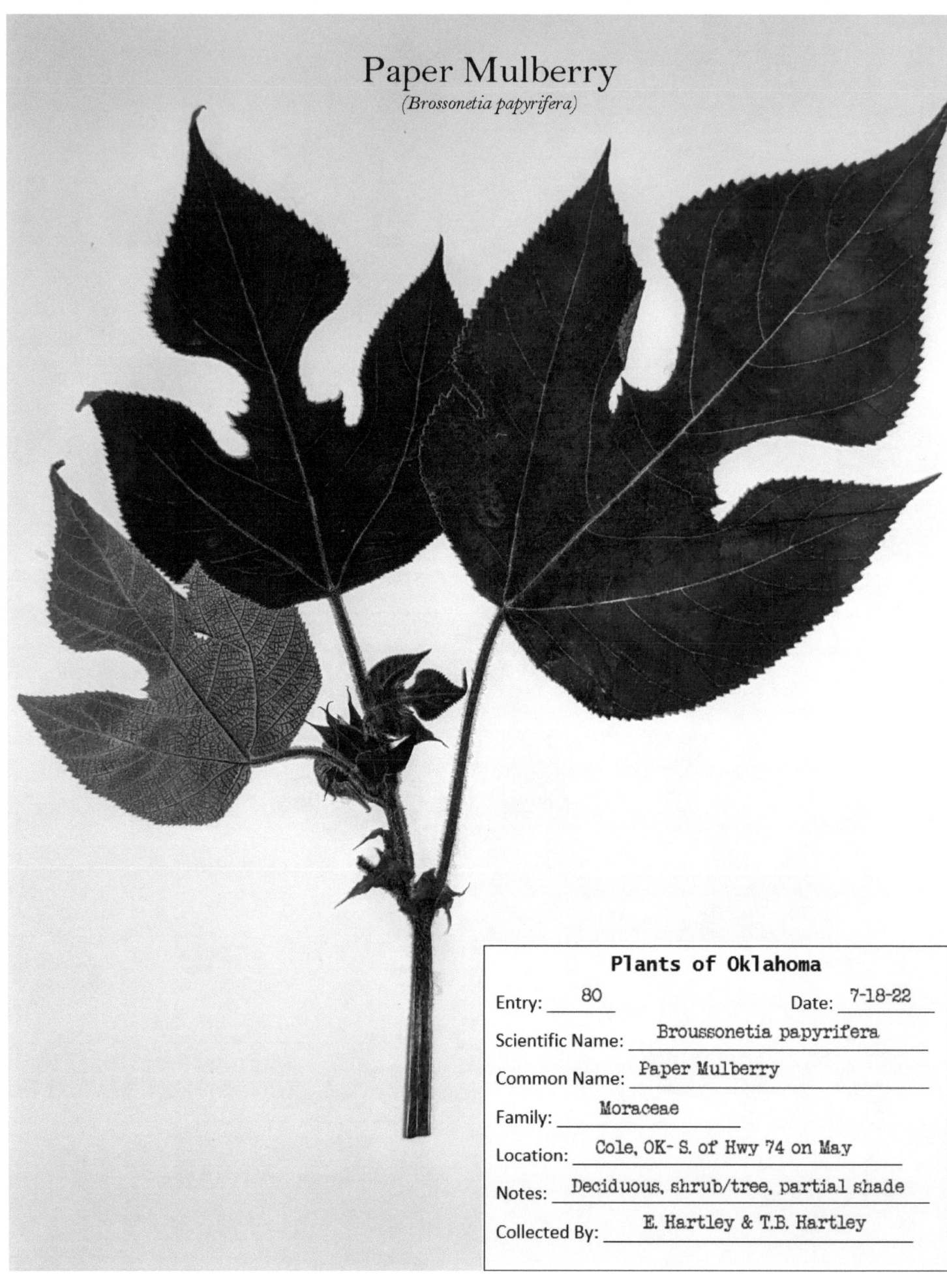

Plants of Oklahoma

Entry: 80
Date: 7-18-22
Scientific Name: Broussonetia papyrifera
Common Name: Paper Mulberry
Family: Moraceae
Location: Cole, OK- S. of Hwy 74 on May
Notes: Deciduous, shrub/tree, partial shade
Collected By: E. Hartley & T.B. Hartley

White Mulberry
(Morus alba)

Plants of Oklahoma

Entry: 10 Date: 6-30-22

Scientific Name: Morus alba

Common Name: White Mulberry

Family: Moraceae

Location: Cole, OK- S. of Hwy 74 on May

Notes: Deciduous, edible fruit & leaves

Collected By: E. Hartley & T.B. Hartley

Red Mulberry
(Morus rubra)

Plants of Oklahoma

Entry: 48 Date: 7-15-22
Scientific Name: Morus rubra
Common Name: Red Mulberry
Family: Moraceae
Location: Walnut Creek, Washington, OK
Notes: Deciduous, edible red/purple fruit
Collected By: E. Hartley & T.B. Hartley

Osage Orange
(Maclura pomifera)

Plants of Oklahoma

Entry: 89   Date: 7-24-22
Scientific Name: Maclura pomifera
Common Name: Osage Orange, Bois D'Arc
Family: Moraceae
Location: 240th & Penn, Cole, OK
Notes: Deciduous, thorny, large green fruit
Collected By: E. Hartley & T.B. Hartley

Trumpet Vine
(Campsis radicans)

Plants of Oklahoma

Entry: 86
Date: 7-20-22
Scientific Name: Campsis radicans
Common Name: Trumpet Vine
Family: Bignoniaceae
Location: Cole, OK- S. of Hwy 74 on May
Notes: Perennial vine, orange flowers, full sun
Collected By: E. Hartley & T.B. Hartley

Southern Catalpa
(Catalpa bignonioides)

Plants of Oklahoma

Entry: 93　　　　　　　　　Date: 8-1-22
Scientific Name: Catalpa bignonioides
Common Name: Southern/Common Catalpa
Family: Bignoniaceae
Location: HWY 77 & Maguire, Noble, OK
Notes: Deciduous tree, long bean pods,
Collected By: E. Hartley & T.B. Hartley

Snow-on-the-Mountain
(Euphorbia marginata)

Plants of Oklahoma

Entry: 97 Date: 7-25-22
Scientific Name: Euphorbia marginata
Common Name: Snow-on-the-mountain
Family: Euphorbiaceae
Location: 250th & May, Cole, OK
Notes: Annual, white flower blooms in Aug.
Collected By: E. Hartley & T.B. Hartley

Hogwort
(Croton capitatus)

Plants of Oklahoma

Entry: 43 Date: 7-12-22
Scientific Name: Croton capitatus
Common Name: Hogwort, Woolly Croton
Family: Euphorbiaceae
Location: I-35 at Goldsby, OK
Notes: Annual, dry gravely soil, full sun
Collected By: E. Hartley & T.B. Hartley

Texas Bullnettle
(Cnidoscolus texanus)

Plants of Oklahoma

Entry: 101 Date: 7-24-22
Scientific Name: Cnidoscolus texanus
Common Name: Texas Bullnettle
Family: Euphorbiaceae
Location: 240th & Penn, Cole, OK
Notes: Perennial, stinging hairs, white flower
Collected By: E. Hartley & T.B. Hartley

Prostrate Spurge
(Euphorbia maculata)

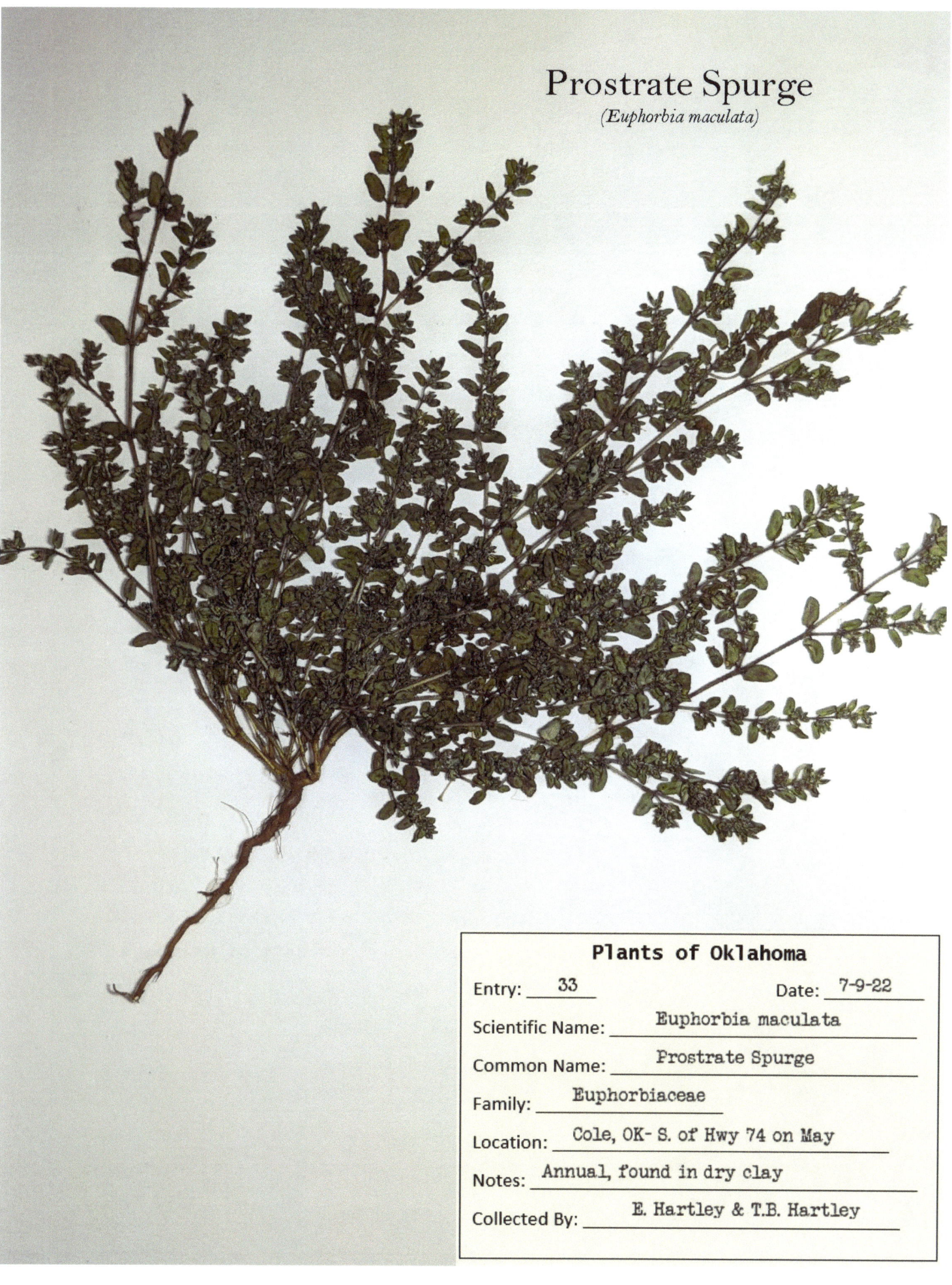

Plants of Oklahoma

Entry: 33 Date: 7-9-22
Scientific Name: Euphorbia maculata
Common Name: Prostrate Spurge
Family: Euphorbiaceae
Location: Cole, OK- S. of Hwy 74 on May
Notes: Annual, found in dry clay
Collected By: E. Hartley & T.B. Hartley

Green Foxtail
(Setaria viridis)

Plants of Oklahoma

Entry: 47 Date: 7-15-22
Scientific Name: Setaria viridis
Common Name: Green Foxtail
Family: Poaceae
Location: Hwy 9, Newcastle, OK
Notes: Annual grass, dry clay, full sun
Collected By: E. Hartley & T.B. Hartley

Bermuda Grass
(Cynodon dactylon)

Plants of Oklahoma

Entry: 52 Date: 7-14-22
Scientific Name: Cynodon dactylon
Common Name: Bermuda Grass
Family: Poaceae
Location: Cole, OK- S. of Hwy 74 on May
Notes: Grass, perennial, common turfgrass
Collected By: E. Hartley & T.B. Hartley

Little Bluestem
(Schizachyrium scoparium)

Plants of Oklahoma

Entry: 70 Date: 7-16-22
Scientific Name: Schizachyrium scoparium
Common Name: Little Bluestem
Family: Poaceae
Location: Cole, OK- S. of Hwy 74 on May
Notes: Perennial, mounding grass, dry sand
Collected By: E. Hartley & T.B. Hartley

Peach
(Prunus persica)

Plants of Oklahoma

Entry: 40 Date: 7-4-22
Scientific Name: Prunus persica
Common Name: Peach
Family: Rosaceae
Location: Cole, OK- S. of Hwy 74 on May
Notes: Deciduous, growing near water
Collected By: E. Hartley & T.B. Hartley

Mexican Plum
(Prunus mexicana)

Plants of Oklahoma

Entry: 100 Date: 8-2-22
Scientific Name: Prunus mexicana
Common Name: Mexican Plum
Family: Rosaceae
Location: 250th & May, Cole, OK
Notes: Deciduous, small green/purple fruit
Collected By: E. Hartley & T.B. Hartley

Sand Plum
(Prunus angustifolia)

Plants of Oklahoma

Entry: 13 Date: 7-2-22

Scientific Name: Prunus angustifolia

Common Name: Sand Plum, Chickasaw Plum

Family: Rosaceae

Location: Cole, OK- S. of Hwy 74 on May

Notes: Deciduous, small edible fruit

Collected By: E. Hartley & T.B. Hartley

St. Lucie Cherry
(Prunus mahaleb)

Plants of Oklahoma

Entry: 39   Date: 7-9-22

Scientific Name: Prunus mahaleb

Common Name: St. Lucie Cherry

Family: Rosaceae

Location: Cole, OK- S. of Hwy 74 on May

Notes: Deciduous, growing in understory

Collected By: E. Hartley & T.B. Hartley

Japanese Rose
(Rosa multiflora)

Plants of Oklahoma

Entry: 75 Date: 7-15-22
Scientific Name: Rosa multiflora
Common Name: Japanese Rose
Family: Rosaceae
Location: 250th & May, Cole, OK
Notes: Perennial shrub, white flowers, invasive
Collected By: E. Hartlev & T.B. Hartlev

White Avens
(Geum canadense)

Plants of Oklahoma

Entry: 36 Date: 7-5-22
Scientific Name: Geum canadense
Common Name: White Avens
Family: Rosaceae
Location: Cole, OK- S. of Hwy 74 on May
Notes: Perennial, dry clay soil, partial shade
Collected By: E. Hartley & T.B. Hartley

Indian Blanket
(Gaillardia pulchella)

Plants of Oklahoma

Entry: 71 Date: 7-18-22
Scientific Name: Gaillardia pulchella
Common Name: Indian Blanket
Family: Asteraceae
Location: 240th & Penn, Cole, OK
Notes: Annual flower, sandy dry soil
Collected By: E. Hartley & T.B. Hartley

Bitter Sneezeweed
(Helenium amarum)

Plants of Oklahoma

Entry: 19 Date: 7-2-22
Scientific Name: Helenium amarum
Common Name: Bitter Sneezeweed
Family: Asteraceae
Location: Near I-35, Norman, OK
Notes: Annual, native, mildly toxic
Collected By: E. Hartley & T.B. Hartley

Hairy Goldenaster
(Heterotheca villosa)

Plants of Oklahoma

Entry: 61 Date: 7-17-22
Scientific Name: Heterotheca villosa
Common Name: Hairy False Goldenaster
Family: Asteraceae
Location: I-35 at Goldsby, OK
Notes: Perennial, yellow flower, sandy soil
Collected By: E. Hartley & T.B. Hartley

Curlycup Gumweed
(Grindelia squarrosa)

Plants of Oklahoma

Entry: 106 Date: 8-8-22
Scientific Name: Grindelia squarrosa
Common Name: Curlycup Gumweed
Family: Asteraceae
Location: 250th & May, Cole, OK
Notes: Biennial, yellow flower, sandy loam
Collected By: E. Hartley & T.B. Hartley

Black-eyed Susan
(Rudbeckia hirta)

Plants of Oklahoma

Entry: 20 Date: 7-2-22
Scientific Name: Rudbeckia hirta
Common Name: Black-eyed Susan
Family: Asteraceae
Location: Cole, OK- S. of Hwy 74 on May
Notes: Annual, found in dry pasture
Collected By: E. Hartley & T.B. Hartley

Common Sunflower
(Helianthus annuus)

Milk Thistle
(Silybum marianum)

Plants of Oklahoma

Entry: 18 Date: 7-3-22
Scientific Name: Silybum marianum
Common Name: Milk Thistle
Family: Asteraceae
Location: Roadside, McClain County, OK
Notes: Biennial, purple flower
Collected By: E. Hartley & T.B. Hartley

Baldwin's Ironweed
(Vernonia baldwinii)

Plants of Oklahoma

Entry: 63 Date: 7-15-22

Scientific Name: Vernonia baldwinii

Common Name: Baldwin's Ironweed

Family: Asteraceae

Location: HWY 62 & Main, Blanchard, OK

Notes: Perennial, native, dry sandy soil

Collected By: E. Hartley & T.B. Hartley

Giant Ragweed
(Ambrosia trifida)

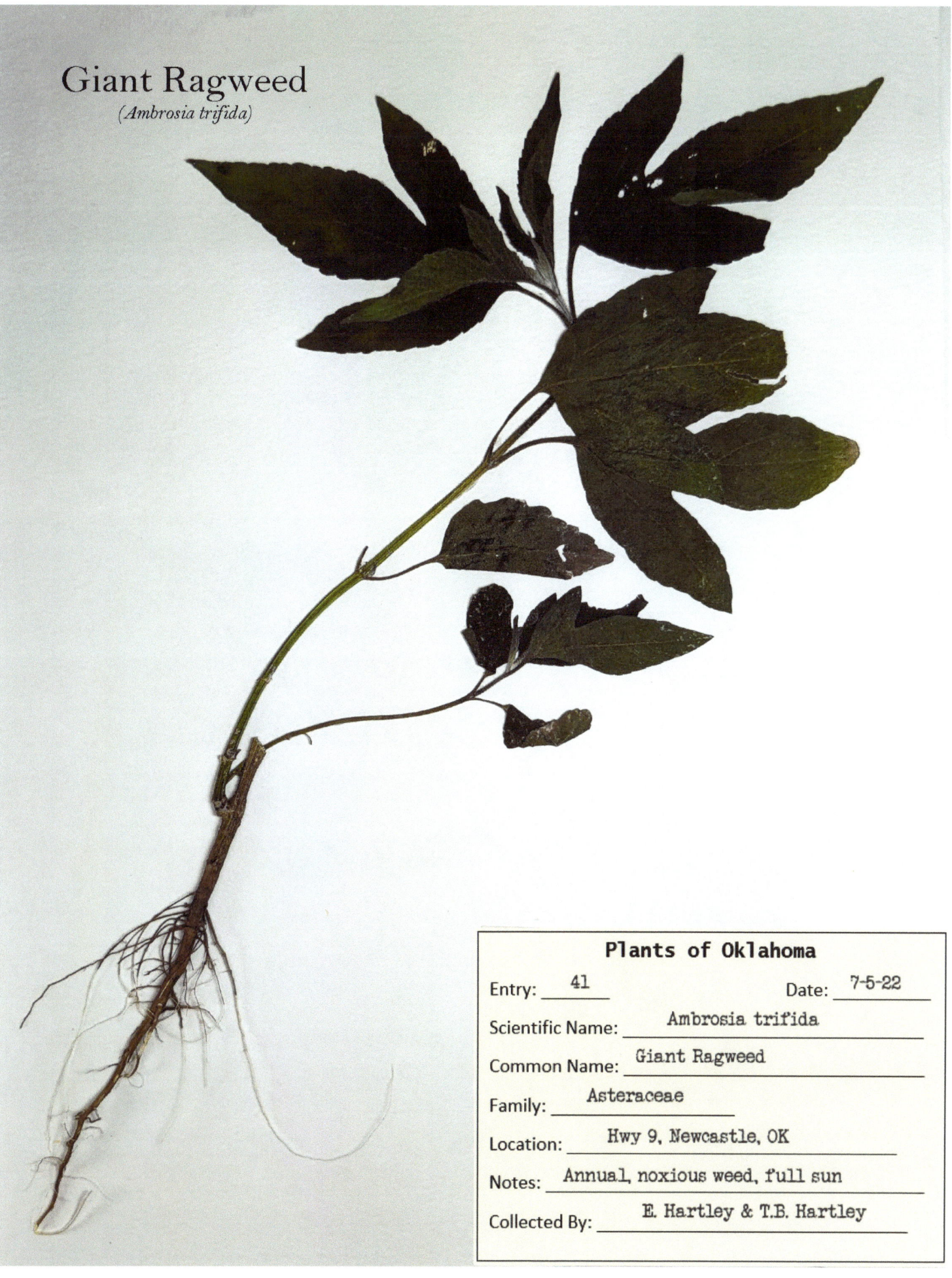

Plants of Oklahoma

Entry: 41 Date: 7-5-22
Scientific Name: Ambrosia trifida
Common Name: Giant Ragweed
Family: Asteraceae
Location: Hwy 9, Newcastle, OK
Notes: Annual, noxious weed, full sun
Collected By: E. Hartley & T.B. Hartley

Western Ragweed
(Ambrosia psilostachya)

Plants of Oklahoma

Entry: 73 Date: 7-17-22
Scientific Name: Ambrosia psilostachya
Common Name: Western Ragweed
Family: Asteraceae
Location: I-35 at Goldsby, OK
Notes: Perennial, dry sandy soil, full sun
Collected By: E. Hartley & T.B. Hartley

Horseweed
(Erigeron canadensis)

White Sagebrush
(Artemisia ludoviciana)

Plants of Oklahoma

Entry: 74 Date: 7-18-22
Scientific Name: Artemisia ludoviciana
Common Name: White Sagebrush
Family: Asteraceae
Location: Cole, OK- S. of Hwy 74 on May
Notes: Perennial, dry clay soil, full sun
Collected By: E. Hartley & T.B. Hartley

Alfalfa
(Medicago sativa)

Plants of Oklahoma

Entry: 59 Date: 7-15-22
Scientific Name: Medicago sativa
Common Name: Alfalfa
Family: Fabaceae
Location: 240th & Penn, Cole, OK
Notes: Perennial, common legume crop
Collected By: E. Hartley & T.B. Hartley

White Prairie Clover
(Dalea candida)

Plants of Oklahoma

Entry: 58 Date: 7-14-22
Scientific Name: Dalea candida
Common Name: White Prairie Clover
Family: Fabaceae
Location: Cole, OK- S. of Hwy 74 on May
Notes: Perennial, native, dry loamy soil
Collected By: E. Hartley & T.B. Hartley

Nine-Anther Prairie Clover
(Dalea enneandra)

Plants of Oklahoma

Entry: 56   Date: 7-12-22

Scientific Name: Dalea enneandra

Common Name: Nine-Anther Prairie Clover

Family: Fabaceae

Location: HWY 37 & Penn, Moore, OK

Notes: Perennial, leguminous, native

Collected By: E. Hartley & T.B. Hartley

Partridge Pea
(Chamaecrista fasciculata)

Plants of Oklahoma

Entry: 54 Date: 7-15-22
Scientific Name: Chamaecrista fasciculata
Common Name: Partridge Pea
Family: Fabaceae
Location: Walnut Creek Bridge, Washington, OK
Notes: Annual, leguminous, dry loam, full sun
Collected By: E. Hartley & T.B. Hartley

Illinois Bundleflower
(Desmanthus dioicus)

Plants of Oklahoma

Entry: 62
Date: 7-18-22
Scientific Name: Desmanthus dioicus
Common Name: Illinois Bundleflower
Family: Fabaceae
Location: HWY 37 & Penn, Moore, OK
Notes: Perennial, leguminous, dry clay soil
Collected By: E. Hartley & T.B. Hartley

Leadplant
(Amorpha canescens)

Plants of Oklahoma

Entry: 67 Date: 7-17-22
Scientific Name: Amorpha canescens
Common Name: Leadplant
Family: Fabaceae
Location: I-35 at Goldsby, OK
Notes: Perennial, purple flower, sandy soil
Collected By: E. Hartley & T.B. Hartley

Redbud
(Cercis canadensis)

Plants of Oklahoma

Entry: 6 Date: 6-30-22
Scientific Name: Cercis canadensis
Common Name: Redbud
Family: Fabaceae
Location: Cole, OK- S. of Hwy 74 on May
Notes: Deciduous, found under tree canopy
Collected By: E. Hartley & T.B. Hartley

Kentucky Coffee Tree
(Gymnocladus dioicus)

Plants of Oklahoma

Entry: 68 Date: 7-18-22
Scientific Name: Gymnocladus dioicus
Common Name: Kentucky Coffee Tree
Family: Fabaceae
Location: Cole, OK- S. of Hwy 74 on May
Notes: Deciduous, large seed pods, sandy soil
Collected By: E. Hartley & T.B. Hartley

Black Locust
(Robinia pseudoacacia)

Plants of Oklahoma

Entry: 15 Date: 7-2-22
Scientific Name: Robinia pseudoacacia
Common Name: Black Locust
Family: Fabaceae
Location: Cole, OK- S. of Hwy 74 on May
Notes: Deciduous, leguminous, toxic
Collected By: E. Hartley & T.B. Hartley

Honey Locust
(Gleditsia triacanthos)

Plants of Oklahoma

Entry: 104 Date: 8-7-22
Scientific Name: Gleditsia triacanthos
Common Name: Honey Locust
Family: Fabaceae
Location: Downtown Purcell, OK
Notes: Deciduous tree, large thorns, seed pods
Collected By: E. Hartley & T.B. Hartley

Mimosa
(Albizia julibrissin)

Plants of Oklahoma

Entry: 5 Date: 6-30-22
Scientific Name: Albizia julibrissin
Common Name: Mimosa, Persian silk tree
Family: Fabaceae
Location: Cole, OK- S. of Hwy 74 on May
Notes: Deciduous, found in moist soil, shade
Collected By: E. Hartley & T.B. Hartley

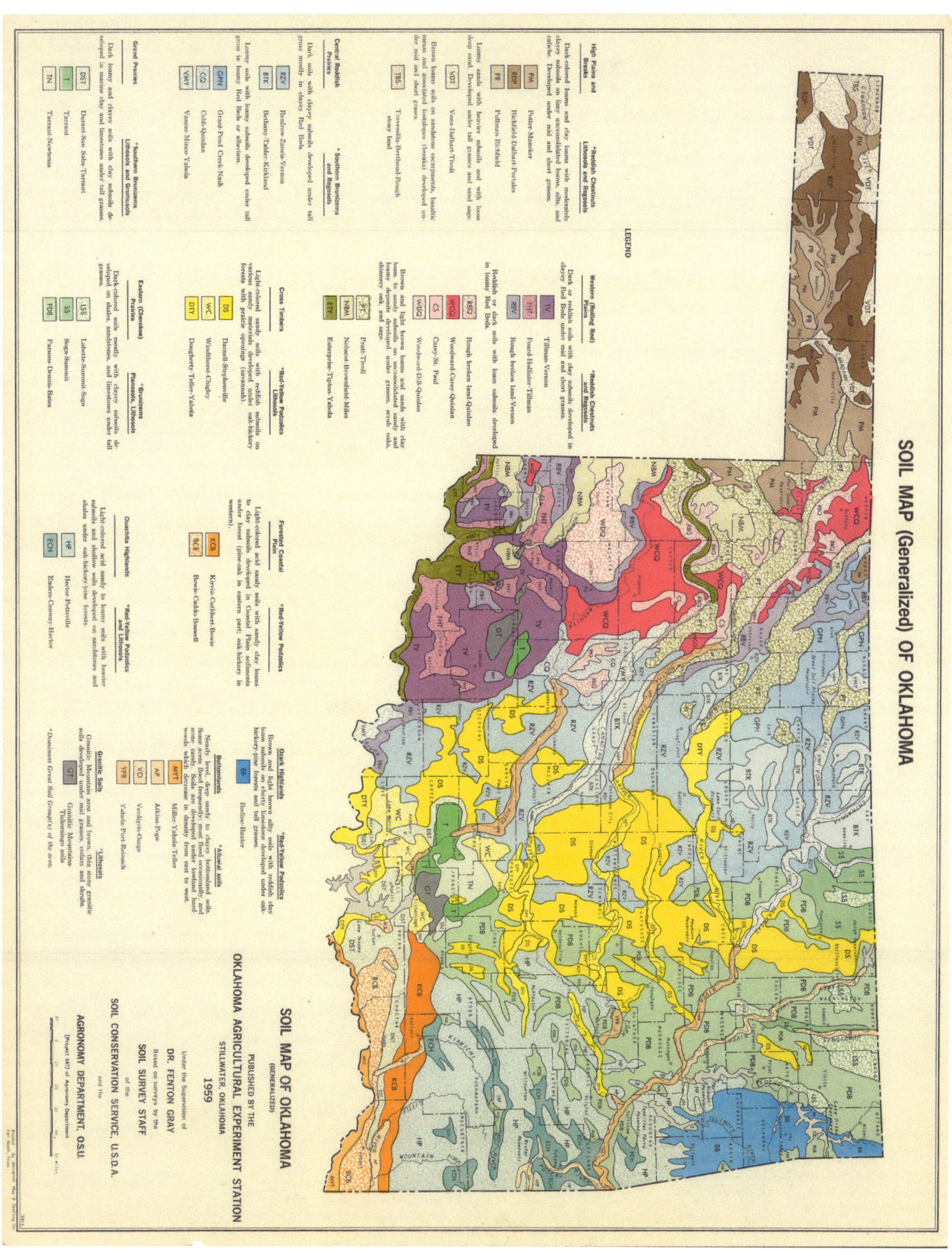

REFERENCES

Fernald, M.L. *Gray's Manual of Botany*, 8th Edition American Book Company, New York, 1950.

Heberling, J. Mason; Isaac, Bonnie L. (2017). "Herbarium specimens as exaptation's: New uses for old collections". American Journal of Botany. 104 (7): 963–965.

Irving, Washington. A Tour on the Prairies, Vol 6 New York, John W. Lovell Company, 1835.

McCoy, Doyle. *Roadside Wild Fruits of Oklahoma*. Norman, University of Oklahoma Press, 1980.

McCoy, Doyle. *Roadside Trees and Shrubs of Oklahoma*. Norman, University of Oklahoma Press, 1981.

Phillips, George R. Gibbs, Frank J. Mattoon, Wilbur R. *Forest Trees of Oklahoma,* Oklahoma City, Oklahoma Forestry Service, 2015.

Smithsonian Institution, *Guide to Herbarium Specimens for Seeds of Success,* https://www.blm.gov/sites/default/files/docs/2021-10/Guide%20to%20Herbarium%20Specimens%20for%20SOS.pdf. Accessed 8/19/2022

TORCH, *The Texas Oklahoma Regional Consortium of Herbaria,* https://portal.torcherbaria.org/portal/index.php, Accessed 8/19/2022.

Williams, John E. *Atlas of the Woody Plants of Oklahoma*. Norman, Oklahoma Biological Survey, 1978.

Woods, A.J., Omernik, J.M., Butler, D.R., Ford, J.G., Henley, J.E., Hoagland, B.W., Arndt, D.S., and Moran, B.C., 2005, Ecoregions of Oklahoma (color poster with map, descriptive text, summary tables, and photographs): Reston, Virginia, U.S. Geological Survey (map scale 1:1,250,000).

INDEX

Alfalfa: 90
American Elm: 13
American Persimmon: 19
American Sycamore: 11
Bald Cypress: 3
Baldwin's Ironweed: 85
Bermuda Grass: 70
Bitter Sneezeweed: 79
Black Haw, Gum Bully: 22
Black Locust: 99
Black Willow: 50
Black-eyed Susan: 82
Blackjack Oak: 6
Box Elder (Maple): 46
Buffalo Gourd, Stinking Gourd: 28
Buffalo-Bur, Spiny Nightshade: 53
Butterfly Weed: 45
Carolina Horsenettle: 52
Chinese Elm: 15
Chinese Pistache: 43
Chinese Privet: 40
Common Milkweed: 44
Common Pepperweed: 26
Common Sunflower: 83
Coralberry: 36
Crabgrass: 68
Crape Myrtle: 21
Curly Dock: 25
Curlycup Gumweed: 81
Dwarf Chinkapin Oak: 95
Eastern Cottonwood: 49
Eastern Poison Ivy: 42
Eastern Red Cedar: 2
Elderberry: 20
Empress Tree: 18
Field Bindweed
Giant Ragweed: 33
Graybark Grape, Winter Grape: 55
Great Mullein: 34
Green Foxtail: 67
Hairy False Goldenaster: 80
Heartleaf Peppervine: 56
Hoary Vervain: 31
Hogwort, Woolly Croton: 64
Holly: 23
Honey Locust: 100
Honeysuckle: 37
Horseweed: 88
Illinois Bundleflower: 95
Indian Blanket: 78
Japanese Privet: 41

Japanese Rose: 76
Johnson Grass: 69
Kentucky Coffee Tree: 98
Lamb's Quarters: 38
Leadplant: 96
Lemon Beebalm: 35
Little Bluestem: 71
Mexican Plum: 73
Milk Thistle: 84
Mimosa, Persian Silk Tree: 101
Nine-Anther Prairie Clover: 93
Osage Orange, Bois D'Arc: 60
Palmer's amaranth, Pigweed: 39
Paper Mulberry: 57
Partridge Pea: 94
Peach: 72
Pecan: 51
Pokeweed, Poke Sallet: 29
Post Oak: 5
Prostrate Spurge: 66
Purple Prairie Clover: 91
Red Maple: 47
Red Mulberry: 59
Redbud: 97
Roughleaf Dogwood: 24
Roundleaf Greenbriar: 30
Sand Plum, Chickasaw Plum: 74
Shortleaf Pine: 4
Shumard Oak: 9
Siberian Elm: 14
Silver Maple: 48
Snow-on-the-Mountain: 63
Southern Magnolia: 17
Southern/Common Catalpa: 62
St. Lucie Cherry: 75
Sugarberry: 12
Texas Bullnettle: 65
Tree of Heaven: 16
Trumpet Vine: 61
Virginia Creeper: 54
Water Oak: 10
Western Ragweed: 87
White Avens: 77
White Mulberry: 58
White Prairie Clover: 92
White Prickly Poppy: 27
White Sagebrush: 89
Yellow Nutsedge: 32

Notes/Leaf Press

Notes

Notes

Notes

Notes

ABOUT THE AUTHOR

Eli Hartley is a lifelong resident of Oklahoma. He is a lifelong student and has a degree from the University of Oklahoma in Interdisciplinary Studies. He loves to travel, write, garden and take on more projects than he will ever finish. He currently lives in Cole, OK with his family, three dogs, a few feral cats, the birds, some beehives, and any other animal who needs a place to stay. He is a writer, editor, amateur movie and food critic, and an overall pretty good listener. Also, he likes to frequent thrift stores, walk barefoot, organize things, and occasionally he will sleep outside for apparently no reason at all. Feel free to contact him for literally any reason via his email: eli.l.hartely@gmail.com. Also, if you like maps, check out his other book, *Old Maps of Oklahoma,* available on Amazon.